Java.

2017 Ultimate Beginners Guide to Learn Java Programming (java for dummies, java apps, java for beginners, coding, java apps, hacking, hacking exposed)

Andrew Butler

CONTENTS

Introduction

Java is a powerful programming language with an impressive array of potential applications. It's useful in both client-side desktop applications and servers-side applications that may not be as visible to the majority of users, but is one of the things that have made it so useful to corporate programmers.

If you want to start writing programs, either for your own personal purposes or for use in your business, Java is an excellent language to learn. It's versatile, flexible, complies with networks, and is relatively secure. For a long time there was a perception that Java was inherently slower than C or C++, and while that may have been true in the past, the continuous upgrades that have been made to the Java Runtime Environment have greatly improved performance speed. In the modern world, you could argue the languages are virtually indistinguishable in terms of run time.

The difficulty level of learning Java will vary depending on your experience with using code. It is a relatively high-level language, meaning it's relatively readable for a human (as opposed to assembly level or other low-level languages that are closer to the computer's natural machine code. There are conventions and syntax specifications involved in programming languages that may take some getting used to if you're a newcomer to code, but Java is very learnable, even for the relative beginner.

The chapters that follow in this book will walk you through the basics of programming in Java. The first chapter is more focused on looking at what the language is and why you would use it, with an introduction to programming languages in the second chapter. In the remainder of the book, you'll learn about the different pieces of a Java code and how they inter-related to form the commands your program gives your computer.

With how computerized everything in our world is today, having the ability to write a program for your computer to execute is a powerful skill. The information in this book can help you to take that first step into the exciting world of writing programs using Java.

Chapter 1 – What is Java?

Even if you're not experienced with programming, you're probably familiar with some element of Java. Since its development by Sun Microsystems in 1995, Java has become one of the most popular and widely-used programming languages. It was first used to develop desktop applications, but has since also been applied to network, server, and web-based programs.

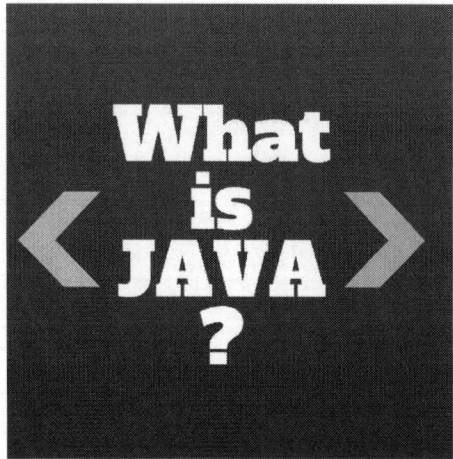

Java refers to both the programming language and the platform used to run the compiled code, which is why you may have some version of Java installed on your system even if you're not a developer. Because it was developed before the era of thumb drives and high speed internet, the original program was broken up into three editions: Standard Edition (SE) for developing client-side applications; Enterprise Edition (EE) for developing server applications; and Micro Edition (ME) which is for making MIDlets and Xlets.

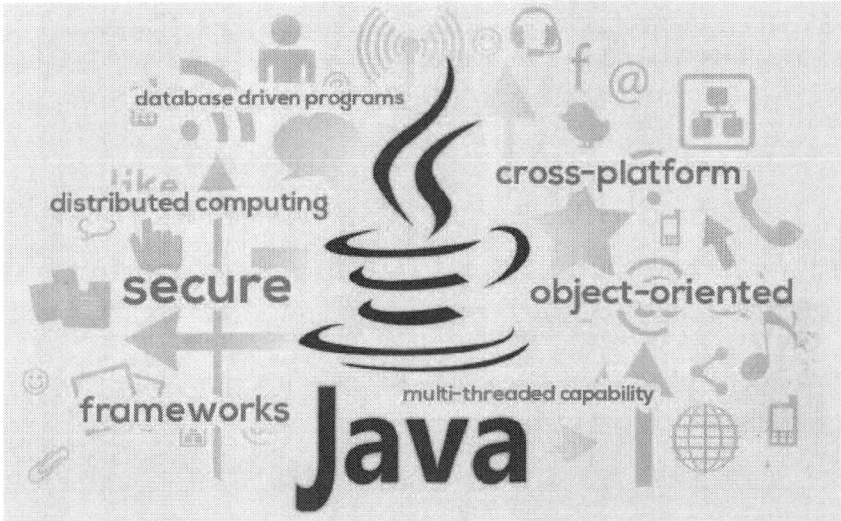

For most people, Java SE is all you need to concern yourself with. This is the edition you'll use to write programs and applications for use on websites, computers, and other devices. As of early 2017, the most recent update of Java SE was Java 8, update 121.

The Java programming language is a general purpose, object-oriented language. The code looks similar to that used in C or C++ but has several key differences that make it easier to use. It doesn't include features like pointers, operator overloading, or multiple implementation inheritance that make C++ confusing to use. The fact that it's object-oriented means you can adapt the program to your need without encountering language constraints. This is in contrast to a structured language like C++ that forces you think about the "state" of an object as different from its "behaviors."

The Java language has other advantages, as well. It is designed to work well with networks, and its extensive network library means it can cope easily with Transmission Control Protocols or Internet Protocols like HTTP or FTP. This simplifies the network connection process. It is also capable of accessing objects across a network using Uniform Resource Locators (URLs).

Since it was designed for use in networked and distributed environments, Java is also a very secure language. It uses public key encryption to prevent nefarious codes or viruses from infiltrating the platform, along with a sandbox-style security model that allows potential breaches to be quickly identified and eliminated.

As a language, Java is both dynamic and multithreaded. Saying it's "dynamic" means it can be selectively upgraded by distributing only the sections of code that have been changed, making it easier to fix bugs and add new features for your users. "Multithreaded" means it can work on two different things at once, for example allowing a Graphical User Interface to remain responsive on one thread while another is waiting for input from a network connection. This vastly improves the performance of programs that accomplish multiple tasks at once.

The platform

Export	PDF	Excel	XML	SOAP	...	

View	Technik	CMD	GUI		Applet	HTML
	Client	Programm	WebStart		Browser	

Controller	Java		Servlet - JSP
			EJB

Model	JDBC (SQL)	ERP	XML
	Database		

This book is mostly concerned with the Java programming language, but you'll be working within the Java platform and it's worth it to take the time to learn a little bit about how the platform turns your code into actions.

The main component of the Java platform is the Java Virtual Machine (JVM), which is the aspect of the platform that translates the instructions and data you program as a developer into language the machine running it can understand. There is also an execution environment on the platform that supports the execution of the JVM.

The general way the platform executes files is relatively simple. First, it loads all the class files being used in the program, designating one as the main class file. After this, the on-board interpreter executes the bytecodes. There is also a verifier on the platform that verifies the loaded class files before executing, allowing you to catch potential errors prior to translation.

The existence of the JVM means that Java executes only indirectly on the underlying program, which is typically an operating system of some kind. This makes it "architecture neutral," meaning the same program can be run equally well on a Mac, Windows, Linux, or Solaris operating system. The JVM translates the code you write into bytecodes, which are easily interpreted by the various systems into the platform-specific instructions used for execution.

Your end user will need to have the Java platform installed on their system in order to run programs written in the Java language. By this point, most computers and devices will come with Java already installed; if they don't, it's free to download online.

Chapter 2 – Introduction to Programming

If you've written code in other programming languages in the past, you can go ahead and skip forward to Chapter 3, which will discuss the set-up of Java on your system. If you haven't, though, you'll probably want to keep reading this chapter to get a sense of exactly what a programming language is, what it does, and how you can use it to create programs.

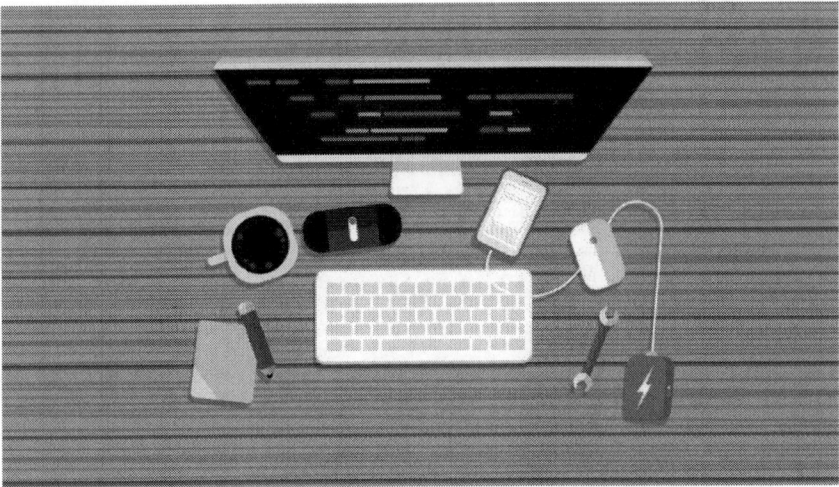

A computer program is basically a sequence of instructions for your computer that tells it how to do a specific task. You can think of it kind of like a recipe you'd use to prepare a meal. It tells your computer what "ingredients" it needs (the objects) and how to prepare them (the statements or expressions you use to manipulate these objects). When your computer carries out the instructions of a program, it is said to "execute" that file.

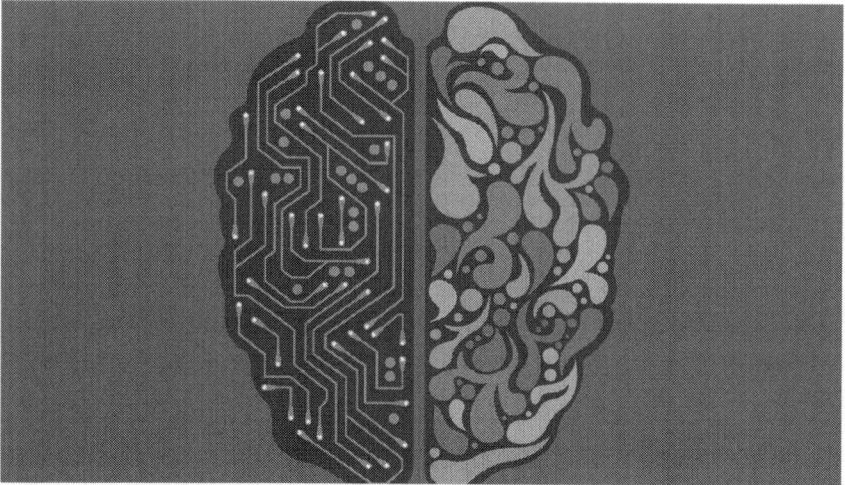

Perhaps the most frustrating thing for a beginning programmer is the fact that you can't make any assumptions about what your computer knows. Using the recipe analogy above, you wouldn't need to tell a human that they need a knife to cut the vegetables for a dish; this is assumed knowledge. Computers, though, can't make these kinds of independent leaps in thought, and must be given explicit instructions from start to finish. While this may take some getting used to, it ultimately also means you get complete control over every action the computer takes as a result of your program.

At its most basic level, a computer is simply a collection of on/off switches. When the switch is off, it has a value of 0; when it's on, it has a value of 1. This collection of ones and zeros that tell the computer which switches should be on and off to accomplish various tasks. The collection of ones and zeros taken together is known as binary.

Every computer comes with a native programming language that is known as "machine code." The machine code is binary-based and is typically specific to the architecture, meaning it would be different between companies and models of computers, and perhaps even from machine to machine.

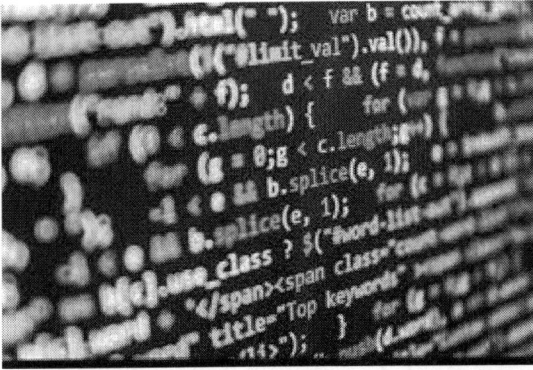

If you wanted to, you could communicate directly with your computer by writing in binary or machine code, but this would be an extremely time consuming process, and it would be very difficult to tell where any errors that come up occurred. Programming languages like Java, BASIC, or C++ were developed to streamline the process of communicating with the computer, letting you write code in a way the human mind can understand. That code is then converted into machine code that's comprehensible for the computer using a platform that's either part of the language's mechanics or installed on the system itself.

There are many different programming languages out there, each of which gives you a unique combination of features and flexibility, and each of which is suitable for a different situation. The sections that follow give a basic overview of the differences between these various programming languages, and what that means for you as a developer.

Compiled vs. interpreted programs

Compile

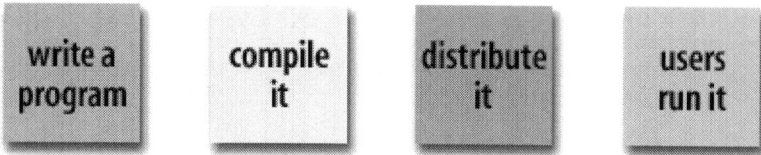

Interpreted

The question of whether a programming language is compiled or interpreted essentially concerns the way the code is translated into the machine code that your computer is able to process. In a compiled language, the code is converted into machine code all at once and then stored in its binary form. In an interpreted language, the code is stored in its human-readable form, and modified into machine code only as it's executed.

To help better understand the difference between these two processes, you can use an analogy of a translated text versus a live interpreter. A compiled language is equivalent to a novel or other text that was written in English but then translated into another language, like Spanish. All the translation work is done before the Spanish speaker picks up the book to read; once the translated book is released, it can be instantly read and understood.

An interpreted program, on the other hand, is more like an English speaker and a Spanish speaker communicating through a translator who speaks both languages. The translation happens in real time, sentence by sentence, with no physical record of the translation. It takes longer at each stage because both speaker and listener will have to pause while the interpreter makes her translation.

Each of these methods has its advantages and disadvantages. In general, an interpreted language is going to be more flexible and adaptable, but is also going to take longer to load since each line of code must be interpreted every time it's called up. Most programming languages are compiled for this reason, but interpreted programs do have their use in certain applications.

Language level

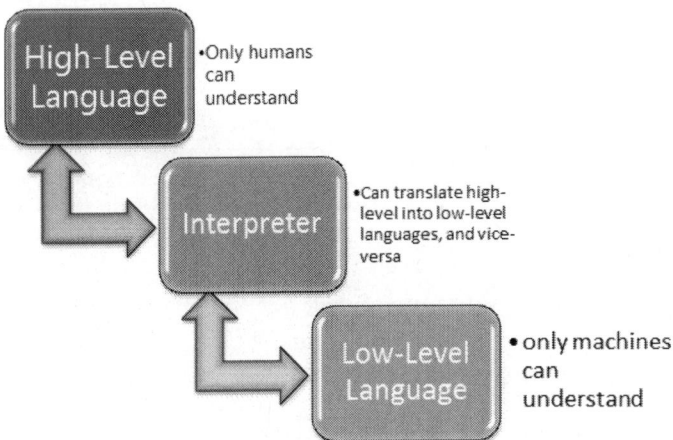

13

The level of a language describes its level of comprehensibility to a computer as opposed to a human being. A low-level language is closer to machine code; a high-level language is closer to a natural language, like English or Spanish.

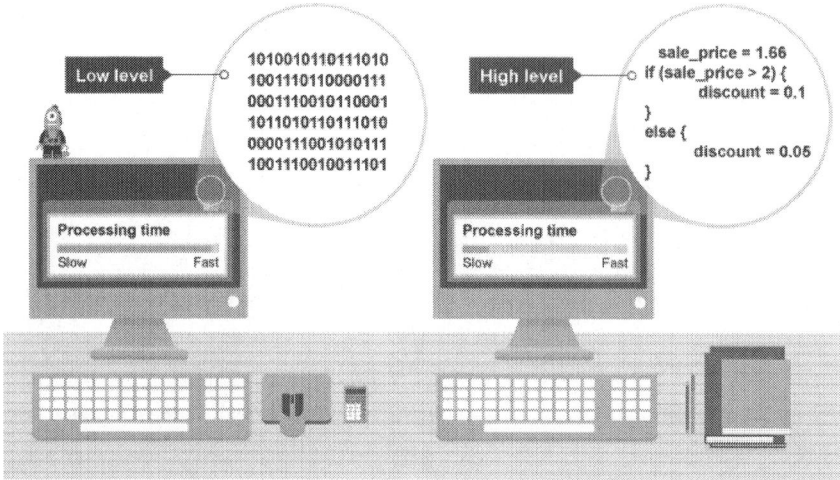

The lowest level language is called assembly language and is essentially a direct translation into alphanumeric characters of the binary instructions executed by the computer. While it's not expressed in ones and zeros, it is still not something the majority of people—even tech savvy ones—would be able to understand. Writing a program in assembly language would be nearly as time consuming and arduous as writing in binary, but it also would allow you to have complete control over every single instruction your computer carries out.

Generally speaking, the higher the level of the language, the less code is required to implement the same feature. Higher level languages are more efficient at expressing instructions, but also draw more on pre-determined processes and actions, giving you less flexibility and control over the implementation of your code.

Implementation

The implementation of a programming language refers to the way the code is translated into machine code and then carried out by the system. The implementation scheme is not built in to the language, and you could theoretically implement any language using any scheme. Practically, however, certain languages are better-suited to specific implementation schemes.

A Compiled Scheme involves the source code being compiled into machine code, which is then interpreted by the CPU. In an Interpreted Scheme, the source code is interpreted line by line as the statements are run. A Mixed Scheme falls part way between these two, with the source code compiled to an intermediate form which is then translated to machine code in real time as the statements are executed.

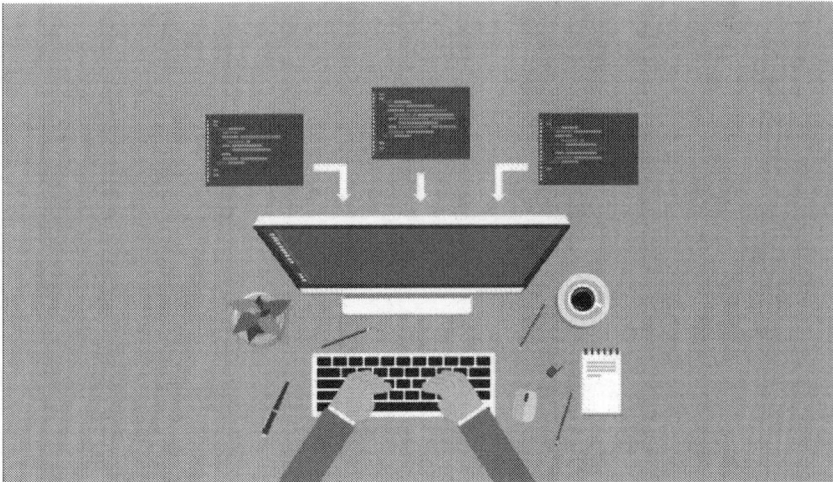

The implementation used by Java is unique compared to most programming languages. It uses a Virtual Machine Scheme. In this scheme, the source code is compiled to a bytecode file that's independent of the platform or architecture. This file is then interpreted by the virtual machine into machine code when it's time for execution.

In general, interpreted languages process more slowly but are more flexible, easier to learn, and faster to develop. As mentioned earlier in the chapter, there is no one best programming language, any more than there's a best implementation. Each language and implementation has its own strengths and is best suited to its own specific purpose.

Chapter 3 – Setting up Java

Now that you have a basic understanding of what Java is (and what programming languages are in general) you're ready to set up the platform on your computer. Again, if you've already accomplished this step or are working on a computer that's been set up for developing, you can skip forward to Chapter 4 to write your first program.

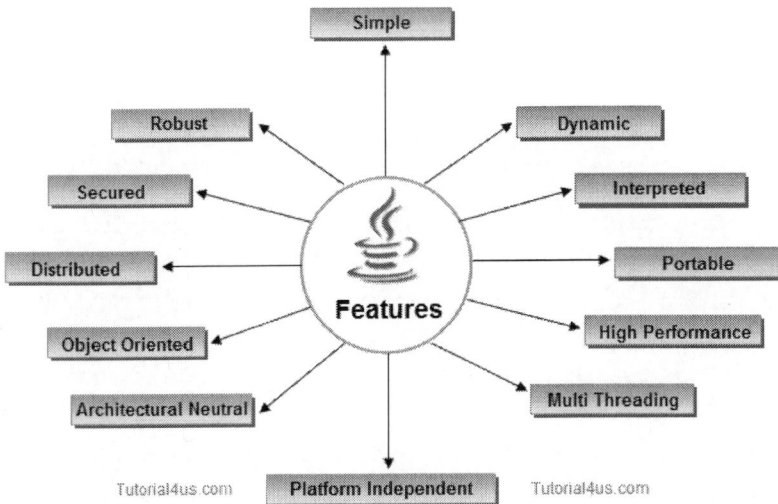

You can run Java on almost any computer. It's available for the standard 64-bit Mac operating systems and both 32-bit and 64-bit Windows systems. It is also available for 64-bit Solaris operating systems, as well as both 32-bit and 64-bit Linux systems.

Your first step should be to download the Java Developer's Kit (JDK) that's appropriate for your operating system. After the file's been downloaded, follow the instructions to install the JDK.

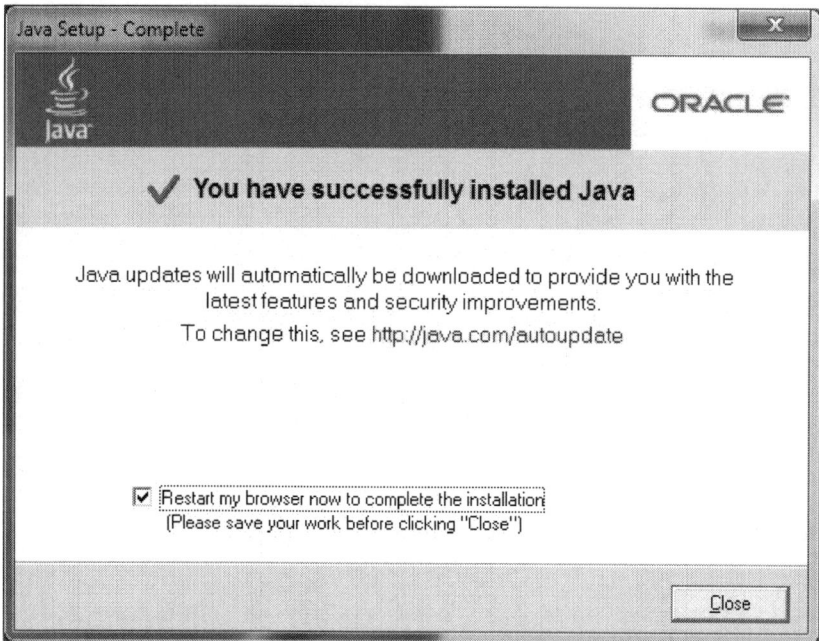

After the program is installed but before you open it up and start writing code, you'll need to update your "path" environment variable to reference the JDK's "bin" subdirectory. This will let you execute the JDK tools from any of the directories in your file system.

If you're not familiar with programming, you may not know about the path. This is basically a system variable that your operating system uses to spot required executables from the Terminal window or command line. Updating your path is relatively simple but does involve going to the behind the scenes areas of your computer.

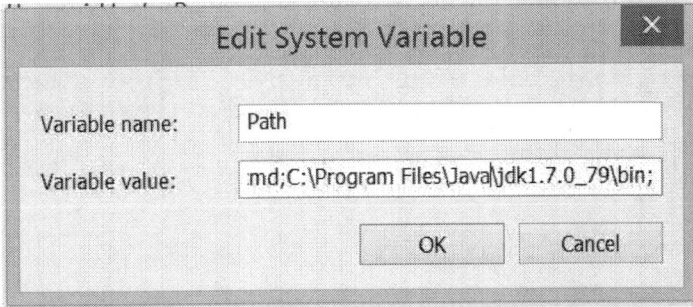

Edit System Variable

Variable name: Path

Variable value: md;C:\Program Files\Java\jdk1.7.0_79\bin;

OK   Cancel

To update your path, you can go to your control panel (on Mac or Windows) or the shell's startup file (on Linux and Solaris systems). Search for "System," then select it and click "Advanced system settings." You should see a section entitled "Environmental Variables," and within that a section labeled "System Variables." This is where you'll find the path environment. If there is one, select it and click "Edit"; if there's not, click "New."

Once you've specified the value of the path in the window that opens up, you can click "Ok" to apply the changes. Once this is completed, you'll have successfully configured your development environment and you'll be ready to start coding your first Java application.

When the JDK is installed on your system it will come with a wide array of files and subdirectories. You can spend some time exploring these on your own, but there are three you'll likely find especially useful.

The "bin" directory contains many of the development tools you'll use on the Java platform. Among these are the Java compiler (javac) and the Java application launcher (java), both of which you'll need to use pretty much any time you write or run a program. The "jre" directory is where you'll go to test your programs. JRE stands for Java Runtime Environment, and this directory contains a copy of this environment that allows you to run any Java program without having to download and install the client-side Java environment. Finally, the "lib" directory is where you'll find the library files that will be utilized by the tools in your Java Developer's Kit.

If you have any issues during the installation or set-up process, you can find help on Oracle's Java SE page. This site is not only helpful for troubleshooting; it's also where you'll find documentation for the most current version of Java SE, along with API references for all the standard class library types and a host of other useful tools and information.

Chapter 4 – Writing your First Program

You've installed your JDK, learned about the virtual machine and its implementation, and have explored the various directories of your Java program. At this point, you're ready to start writing your first program—and the rest of this chapter will walk you through that process, step by step.

First, though, there are a couple more things you should understand about Java. One main thing to start internalizing from the beginning is the fact that both capitalization and spacing matter in Java. This means "Accounts" will be a different class than "accounts," for example, and can be a both useful and a frustrating aspect of the language.

The other knowledge that will be helpful as you're embarking on your first program is to have a basic grip on some terms and definitions. This program could also be described as an "application," which in the context of Java is any code that contains, at the least, a single class that declares a main method.

Of course, this definition is relatively useless to you until you have a few more terms in hand. A "class" is basically a placeholder for the storage location of the data item in question which allows you to then declare methods concerning that data.

A "method" is a block of code that processes the data put into it and returns an output. If there's no output for the action in question, you can also put the word "void" in front of the name of the method.

When you write code in Java, you'll start by declaring the class; all of the information in that class is then enclosed in braces (the { and } symbols on your keyboard). The next thing in the code will be the method, which will also have a body of code following it enclosed in braces. Typically the method is tabbed over once to help guide the eye down the code.

Hello, world

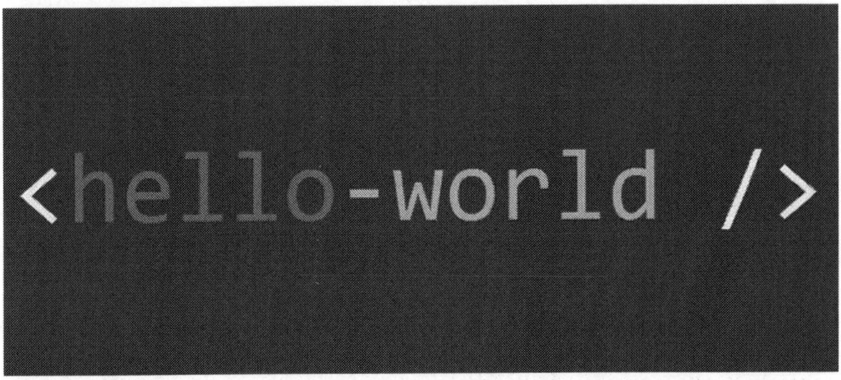

Whenever you start to learn a new programming language, the very first program you're likely to learn is one that simply outputs the phrase "hello, world" onto your screen. In this case, you'll be naming the program HelloWorld.java, and the code for it will look like this:

```
class HelloWorld

{

    public static void main(String[] args)

    {

        System.out.print1n("hello, world");

    }

}
```

This is a very short piece of code, obviously, but it gives you a general introduction to everything you need to know about Java. Notice the class designation at the top, which is named "HelloWorld" just like the file name.

The method can be found on the second line of text, and is the portion that reads "main ()"; all the other text on that line is to modify this method. "Public" and "static" tell the system what permissions the program has and allows the Java runtime environment to access them. "Args" is short for "arguments" and is the name of the array being referenced. Strings will be explained more in the next chapter.

The execution of the method, meanwhile, is found on the third line of text, which is the actual statement being executed by the JVM. In this case, it's telling the system to print the phrase "hello, world" to the standard output stream, which is typically set as the command line.

Once you've saved this code as the file HelloWorld.java, go to the command line and type "javac HelloWorld.java." This will set your JDK to compiling the source file. If it compiles without error, you should see a file entitled "HelloWorld.class" in the directory—the executable equivalent of HelloWorld.java.

To run the program, go back to the command line and type "java HelloWorld." Though the file ends with .class, you don't want to include that extension in the command line or you'll get an error message. If you've done everything correctly, you should see the output of "hello, world" appear on the command line.

Hello, world (take 2)

You've successfully completed your first Java program—congratulations! Of course, you don't want to stop there. Before we delve into more specifics about the components of a Java code, let's stop for a moment and play around with this first program a little bit.

First, alter the code in your HelloWorld.java program so that it reads like this:

```
class HelloWorld

{

    public static void main(String[] args)

    {

        System.out.print1n("hello, " + args[o]);

    }

}
```

You'll notice the only change from the first block of code is in the third line of text. The "world" has been deleted from "hello, world"; instead, a new argument command has been added that appends the string from element 1 of the arguments array to the "hello" greeting.

```
public class HelloWorld {
  public static void main(String[] args) {
    System.out.println("Hello World!");
  }
}
```

Compile this code like you did with the first version of HelloWorld.java. Now try to run it the same way you did, by typing "java HelloWorld" into the command line. This time, you won't get the "hello, world" output you got before. Instead, you'll get an error message that tells you the array had an out of bounds exception in the main thread. This is because the addition of "args[0]" leads the program to expect a command-line argument; when it doesn't get one, it can't successfully execute the program.

```
cmd
D:\java>java Hello
Exception in thread "main" java.lang.NoClassDefFoundError: Hello
Caused by: java.lang.ClassNotFoundException: Hello
        at java.net.URLClassLoader$1.run(URLClassLoader.java:200)
        at java.security.AccessController.doPrivileged(Native Method)
        at java.net.URLClassLoader.findClass(URLClassLoader.java:188)
        at java.lang.ClassLoader.loadClass(ClassLoader.java:307)
        at sun.misc.Launcher$AppClassLoader.loadClass(Launcher.java:301)
        at java.lang.ClassLoader.loadClass(ClassLoader.java:252)
        at java.lang.ClassLoader.loadClassInternal(ClassLoader.java:320)
Could not find the main class: Hello.  Program will exit.
```

Error messages in Java shouldn't be seen as points of frustration, but instead should be viewed as helpful aides in perfecting your code. There is a lot of useful information in an error message. Not only does it tell you the nature of the exception (in this case, an array index that's out of bounds) but it tells you the thread where the exception occurred (main, in this case) and even tells you the exact line of code on which the error occurred (the portion in parentheses at the end of the error message; the error is on line 5).

Now go back to the command line. Type "java HelloWorld" like you did before, but before you run it, type your name after this command. If your name is Joe, for example, you should now get an output that says "hello, Joe." You can type whatever you want after the command and the program's output will say hello to it.

The name you type after the command is what is known as a "command line argument." The addition of "args[0]" to the code tells Java that it should look for more information (an argument) before executing the program. This is another very important lesson to learn about Java before we move forward: if you tell the program to do something, it will not function correctly unless it's able to fulfill that command, unless the command is expressed with a Boolean or conditional expression. As is the case with most programming languages, precision is important to get the results you want out of the program.

Chapter 5 – Deconstructing Java Code

One of the best ways to learn a new programming language is not necessarily to try writing a lot of programs yourself but is instead to study the code written by others. These codes may seem incomprehensible at first—especially the more complicated programs—but picking them apart line by line and being able to identify what each specific term accomplishes in the overall code is a great way to gain a deeper understanding of Java, as well as programming in general.

The sections that follow in this chapter are aimed at explaining all the basic building blocks of a Java code. Some of them, like types and identifiers, are present in every single program; others are optional but exceptionally common, to the point you'll rarely find a program without them.

You may find it helpful to look up code written by somebody else and go through it looking for these different code expressions while you're going through this chapter. Seeing the various aspects of Java in action within a pre-made program can be one of the best ways to gain a full understanding of what the code is there to accomplish and how you employ it within the program.

Identifiers

Code entities such as classes and methods have to be named in order to be referenced and utilized by the program. The portion of the source code that gives these items their name is known as the identifier, and is both one of the most important aspects of your code and one of the easiest to pick out when you're first learning.

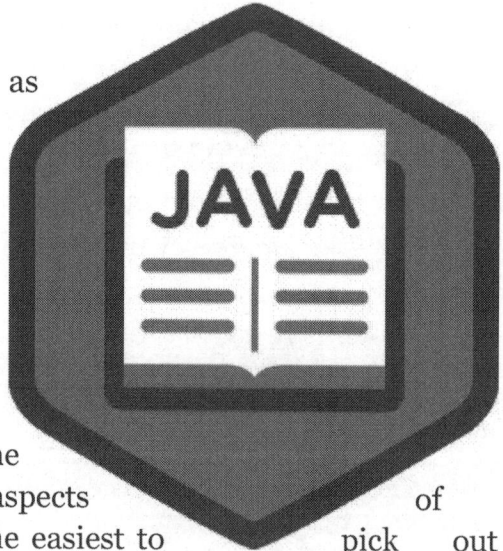

The identifier can utilize the widest range of characters of all the types of code that you'll be using in Java. You can use letters, both English and non-English, which means you can use Greek letters or letters with accents and special symbols. You can also use digits, connecting punctuation (such as underscores and hyphens) and currency symbols, like dollar signs.

Any other symbols can't be used in identifiers, including spaces. You also can't start the identifier with a digit, even though you can use them later in the identifier. The program will also give you an error message if the identifier wraps from one line to the next. Like other things in Java, identifiers are case sensitive, meaning the identifier "Account" will be a different object than "account."

Most words are viable to use as identifiers, but there are some exceptions—mostly words that are used to denote other activities, objects, or other aspects of the program. These are what are called "reserved words," and the compiler will output an error message when it detects these words used outside their usage context.

The full list of reserved words is as follows: abstract, assert,

Boolean, break, byte, case, catch, char, class, const, continue, default, do, double, else, enum, false, final, finally, float, for, goto, if, implements, import, instanceof, int, interface, long, native, new, null, package, private, protected, public, return, short, static, strictfp, super, switch, synchronized, this, throw, throws, transient, true, try, void, volatile, and while.

Types

Every variable, expression, and other value you reference in your program is assigned a type by the computer. The type helps the compiler to detect potential errors with the code at the time of compiling, which lets you address them earlier in the process than if they're not detected until runtime.

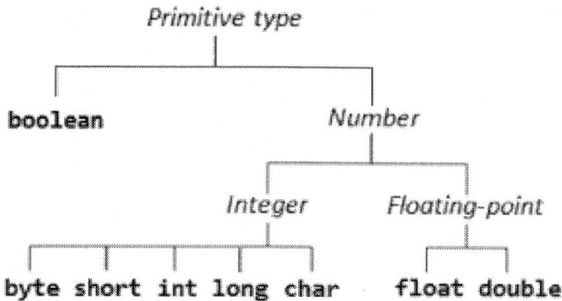

Java deals with a wide variety of different values, from the very large to the very small. Different types are suited to different situations, and will give you different value ranges and compiling times. Classifying the values into different types instead of just using one for all purposes makes it much quicker to complete certain processes when you're working within a limited numerical parameter.

There are three different general types: primitive, reference, and array. All of them are in some way a set of values and how they're represented in the memory, along with the operations for manipulating them. The sections that follow explain each of these in more detail, along with when you'd use them and what they're best suited for.

Primitive types are defined using language and have values that are not objects. Many—but not all—primitive types are sets of numbers, including integer, byte integer, long integer, short integer, floating point, and double precision floating point.

Java Primitive Types

Type	Size	Range	Default
boolean	1 bit	true or false	false
byte	8 bits	[-128, 127]	0
short	16 bits	[-32,768, 32,767]	0
char	16 bits	['\u0000', '\uffff'] or [0, 65535]	'\u0000'
int	32 bits	[-2,147,483,648 to 2,147,483,647]	0
long	64 bits	$[-2^{63}, 2^{63}-1]$	0
float	32 bits	32-bit IEEE 754 floating-point	0.0
double	64 bits	64-bit IEEE 754 floating-point	0.0

The four types with "integer" in their names are all sets of whole numbers. All are stored as two's-complement values. The main difference between them is how wide a range of integers is included in the type.

The **byte integer** type is the smallest, representing integers in 8 bits which gives you a range of values from -128 to 127. When this type is compiled, a bytecode is first generated, which is then converted to an integer value before any operations are performed. This can make it slightly slow to compile. The byte integer type is most useful for smaller values used in an array. It is indicated with the word "byte."

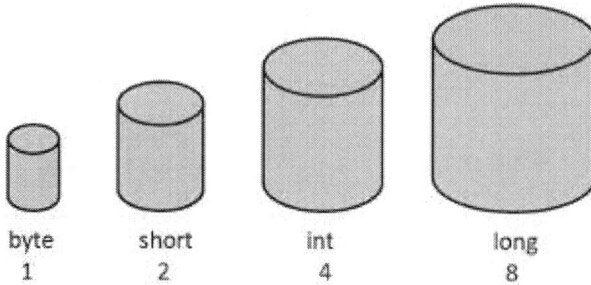

byte 1   short 2   int 4   long 8

The **short integer** type describes integers in 16 bits, giving you a range of values from -32,768 to 32,767. It undergoes a similar compilation process as byte integer types but is slightly faster. It is indicated with the word "short."

The **integer** type describes integers in 32 bits, giving you a range of -2,147,483,648 to 2,147,483,647, and is indicated with the word "integer." Finally, the **long integer** is in 64 bits and has a range of -2^{63} to $2^{63}-1$. It's indicated with the word "long."

If you want to work with fractions or decimals, you have two options: **floating point** type or **double precision floating point** type. Both are represented in IEEE and give you a wide range of values to work with. Floating type is smaller in both space occupied and range of values available, and is most useful in arrays; you can indicate it with the word "float." Double precision floating point type is indicated with the ID "double." Not only does it have a wider range, it's also more precise, giving you 15-17 decimal digits of precision as opposed to the 6-9 with floating precision type.

Finally, there are also two primitive types that don't define numeric values: Boolean and character. **Boolean** type describes true/false values. They aren't manipulated using mathematical operations but instead with AND, OR, and NOT commands. Boolean values are represented as 8 bit integers when stored in an array but are represented as 32 bit integers when they appear in expressions. It is indicated by the word "Boolean."

```java
public static boolean scoresIncreasing(int[] scores)
{
  boolean increasing = true;
  int x = 0;
  while ((increasing == true) && (x < (scores.length - 1)))
  {
    if (scores[x] < scores[x + 1])
    {
      increasing = true;
    }
    else
    {
      increasing = false;
    }
  }
  return increasing;
}
```

The **character** type gives you a range of alphanumeric values, which are indicated with their Unicode number. These character values are represented in the memory as 16 bit integers. The main action you would take on character type values is a classification, which can tell you which values are letters and which are numbers or similar data.

Reference types are types from which objects can be created or referenced. The reference serves as the pointer to the object, whether that's in the form of a memory address or an index in a table of various memory addresses. They're also known as user-defined types because the language users are typically the ones who create them.

References can be kind of confusing if you're not familiar with programming languages until you see them in action. Let's say you're using the reference tool to create a list of flower names. You might write a block of code that looks something like this:

```
class Flower

{

    String name;

    Flower (String flowerName)

    {

        name = flowerName;

    }

    String name()

    {

        return name;

    }

}
```

In the code above, the "name" field stores the names of various flowers in a String (more on those later in the chapter). The method name() returns a flower's name when you run a statement, using a separate code like the one below:

```
Flower = new Flower("tulip");

System.out.print1n(flower.name());
```

Assigning reference types to objects can help you when you're setting up a searchable databank or a program with a large number of objects that you want to be able to sort into more manageable categories.

Array type is a special reference type that denotes an array, which is a region of memory in the program that stores values in ordered slots o equal size. The values within an array are known as elements.

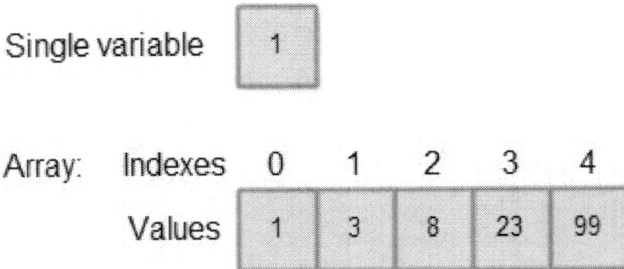

Single variable   [1]

Array: Indexes   0   1   2   3   4
        Values | 1 | 3 | 8 | 23 | 99 |

To set up an array, start by indicating the type of the elements within the array, followed by pairs of square brackets to indicate the dimensions. For example, if you want to have a one dimensional array (also known as a vector) and all the values were integers you would use "int[]" to indicate this.

Multiple sets of brackets denote different kinds of arrays. Two sets of brackets indicates a table (a two dimensional array) and three sets signifies a one dimensional array of two dimensional arrays (a vector of tables). You can also use arrays with strings and use those elements.

The final type you should know about is the **void** type, represented by typing the word "voice." It's used in the method header to indicate that a method doesn't return a value, such as in the HelloWorld.java example from chapter 4.

Literals

Literals are a feature that let you embed values in a source code, essentially the value's character representation. Each primitive type is associated with its own set of literals. Boolean uses true and false; integer and floating point use digits; and character type uses alphanumeric characters placed within single quotes.

Example of Literals in Java					
Integer	5000	0	-5		
Floating-point	6.14	-6.14	.5	0.5	
Character	'a'	'A'	'0'	':'	'_'	')'
Boolean	true	false			
String	"f5java"	"3.14"	"for"	"java"	"int"	"this is a string"

By default, integer literals are assigned to the int type; if you want it represented with the long integer type, suffix a capital L. You can also use non-standard numerical systems. For binary, use the prefix 0b or 0B followed by a sequence of ones and zeros. For hexadecimal, use the prefix 0x or 0X followed by a combination of the characters 0-9, a-f, and A-F. For octal, use the prefix 0, followed by the digits 0-7.

If you want fractions or decimals, the system will default to the double precision floating point type. If you'd rather use floating point type, use the symbol F or f. You can also use this type for exponents.

When you're working with literals, you have to be careful that you don't accidentally trigger a code sequence you didn't intend to. To avoid this, you can use escape sequences. These are representations for characters that can't be expressed literally.

Start an escape sequence with a backslash. If you want to use a literal backslash, you always have to escape sequence it, or the system will think you're starting an escape sequence. You will also need to do this for double and single quotes.

There are specialized escape sequences that you can use as well to get symbols you can't use otherwise. Escape sequences exist to give you a backspace (\b), a form feed (\f), a new line (\n), a carriage return (\r), or a horizontal tab (\t). You can also use a Unicode escape sequence by typing \u followed by the four-digit hexadecimal code for the character you wish to represent.

Since you can't use spaces, long strings of literals can be confusing to read. You can improve the readability by inserting underscore characters between the digits. Don't try to use a leading underscore or the system will think it's an identifier; you'll also get an error message if you try to use a trailing underscore.

Variables

A variable is named space in the memory where a value is stored. Sometimes the variable is the value, as when you're using primitive types; other times it references an object that's stored elsewhere, such as with reference types.

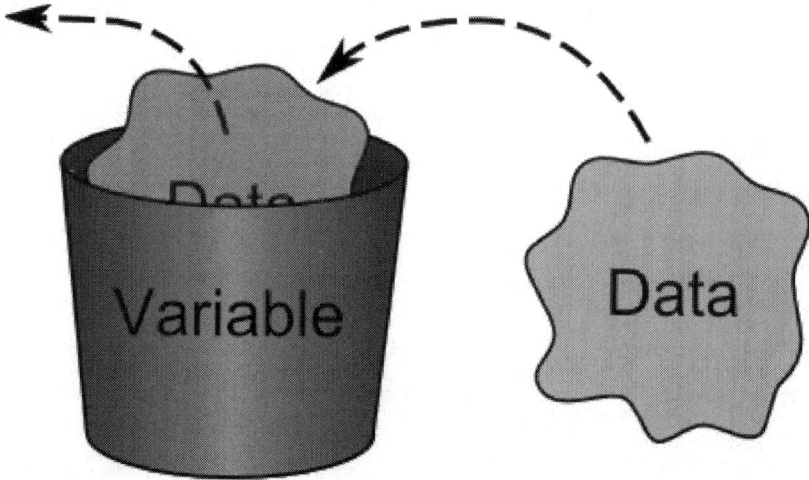

A variable has to be declared because it can be used. It contains, at minimum, a type name, often followed by a sequence of square bracket pairs or by another name. It is terminated with a semicolon. So a variable written "int age;" would declare the variable "age" and its type integer. The variable "char[] text;" declares the variable "text" to be of the character type, with a one dimensional array. The variable "double[] [] temps;" will identify the variable "temps" as a double precision floating point type with a two dimensional array.

A variable also needs to be initialized before use, which you can do as part of the declaration. To initialize a variable, type an equal sign followed by a literal object creation expression that either begins with the word "new" or an array initializer. An array initializer for the temp variable example above might be:

double [] [] temps = { { 34.1, -10.4, 97.2}, {56.0, 88.9, 5.3}}

Within an array initializer, you should include braces around the values. If there are multiple dimensions to the array, each should be in its own nested set of braces. The values within each array should be separated by commas.

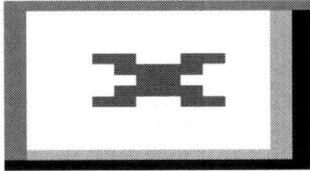

When you're ready to access the variable's value, it only takes a simple line of code to retrieve it. If you're using a reference type you'll have to de-reference the object first. You can de-reference the object by putting a period between the reference variable and the method. Remember the flower name example from the reference type section? It contains a line of code that does this:

System.out.print1n(flower.name());)

Here, the period is de-referencing the "flower" variable from the "name()" method. Once that's done, you're ready to access the variable. Going back to the temps example from earlier in this section, the code to do this would look like this:

System.out.print1n(temps[0][1]);

The numbers in the bracket are how you access the array. If there were only one dimension to the array, you would simply put a 0; for further array dimensions, you would continue to add 1 for each new set of brackets.

You can declare multiple variables in the same declaration as long as they share the same type. Simply separate them with a comma following the type designation. If one of the variables has an array, the brackets can follow that variable name (example: int x, y[], z;). If they all have arrays, you can put the brackets after the type instead (int[] x, y, z;).

Strings

A string is a sequence of characters that is considered together as a single object for the purpose of your program. Each element within the string is separated from the others using double quotes, and you use the string() class designation to create and manipulate the strings used in your program.

Chapter 6 – Expressions

An expression in Java is a combination of literals, methods, variables, and operators that, when evaluated, produces a new value. It is similar to the meaning of expression used in mathematics, and indeed many expressions—especially those with values of an integer or floating type—are various forms of mathematical equations.

There are two main kinds of expressions: Simple expressions

and compound expressions. A simple expression contains a literal, method call, or variable, with an operator. Simple expressions have a type like other variables, which will either be a reference type or some kind of primitive type.

A compound expression is one or more simple expressions which are integrated into a larger expression using an operator. If you're not familiar with this term, an operator a symbol representing the instruction (or sequence of instructions) the program is to follow to manipulate the two simple expressions. Through use of this operand, a new value is produced. Again, you can look to mathematics for an example. In the simple equation 1+2=3, the plus sign is the operand, using the given values of 1 and 2 to create a new value (3). Compound expressions in Java follow this same basic formula.

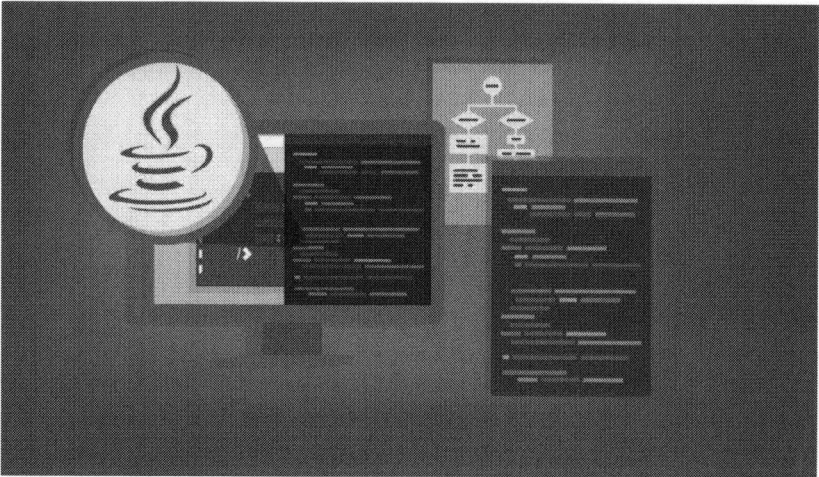

You can combine compound expressions into larger expressions, as well. When you do this within a single expression, Java will follow its pre-programmed rules of precedence to determine the order in which they're processed. This order is as follows:

- Array indexes, member access, and method calls

- Postdecrement or postincrement

- Bitwise complement, logical complement, cast operator, object creation, and unary minus

- Predecrement or preincrement

- Division, multiplication, or remainder

- Addition, subtraction, or string concatenation

- Shifts (left, right, or unsigned)

- Greater than, less than, greater than or equal to, less than or equal to, and type checking

- Equality and inequality

- Bitwise AND or logical AND

- Bitwise exclusive OR and logical exclusive OR

- Bitwise inclusive OR and logical inclusive OR

- Conditional AND

- Conditional exclusive OR and conditional inclusive OR

- Conditional

- Assignment and compound assignment

If there are multiple operands from the same level of precedence in a compound expression, JAVA will process these from left to right through the expression. If you want to violate this order of precedence, you can do so by placing subexpressions within parentheses. All expressions within parentheses are evaluated first—moving left to right—before the program moves on to other operands. You can also nest parentheses if you want to control multiple levels of the evaluation. In the case of nested parentheses, the innermost are evaluated first.

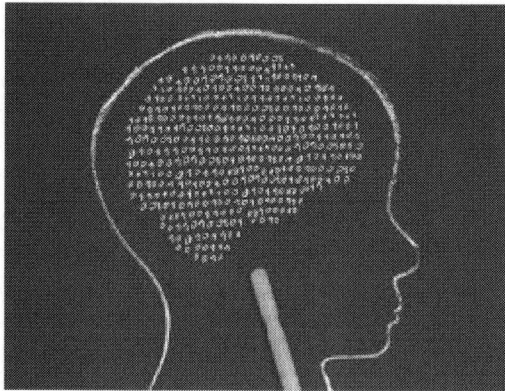

Operators are also classified by the number of expressions that they affect. Operators that only affect a single expression, such as a negative sign before a single value, are known as unary. Operators that affect two expressions are known as binary; that includes the majority of familiar mathematical operators, like addition and subtraction.

Finally, those that affect three expressions are called ternary. This category includes conditional operators.

If the operator comes before the expression, it is known as a prefix operator. If it comes after the expression, it's a postfix operator. Those that fall between two expressions are called infix operators, and are the most common, again including nearly all of the common mathematical operations available to you.

Java gives you a whole host of different operators to work with that allow you to manipulate expressions and values in a wide variety of different ways. Some of these symbols and actions are likely familiar to you, whether or not you've worked with programming languages before, while others are more unique to the coding world. The sections below go into more detail about all of the operators you can use on Java, how to use them in the code, and what effect they have on values.

Additive operators

An additive operator is considered anything that increases or decreases the numeric value of the expression or expressions involved in the equation. The most familiar of these symbols are the two binary operations of addition (+) and subtraction (-), but those aren't your only option.

Several of the additive operators in Java are unary. A postdecrement is indicated by two minus signs at the end of the expression (--). It subtracts 1 from the value of the variable, then stores the result of this subtraction and returns the original variable. You can also do a predecrement, where the double minus signs are placed before the expression. A predecrement will still subtract 1 from the value of the variable and store the result, but it will also return the new variable to you.

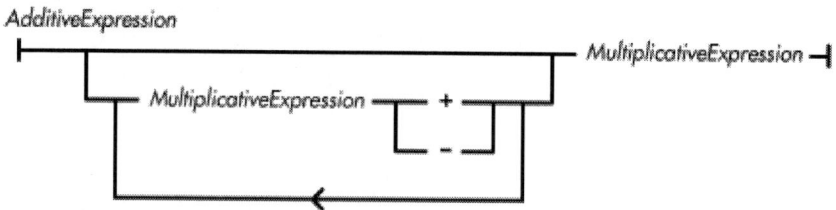

The operators preincrement and postincrement are expressed by a double plus sign (++) and follow the same basic concept: both increase the value by 1 and store this new value, but a postincrement returns the original value to you, while a preincrement returns the new value. These four operators are especially useful when you're trying to run an iteration statement, and their most common use in code is to advance the program to the next iteration.

The final additive operator is called a string concatenation.

Though this term might make it sound like a complicated operation, in truth it's rather simple—this just refers to a form of addition in which at least one of the expressions being altered is of the String type.

Multiplicative operators

Like additive operators, these operations should at least be familiar to you, even if the symbols used to represent them in Java are not. Multiplication is denoted by an asterisk (*) and division is denoted by a slash (/). There is also an operator to get the remainder, represented by a percent sign (%). This will divide one expression by another and return the remainder.

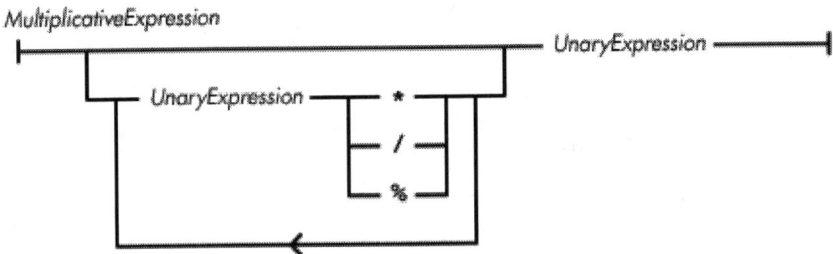

Relational operators

Being able to route the program into different paths depending on where in a range a value falls is a powerful and useful skill, and you can do this using relational operators. There are four to choose from: greater than (>), less than (<), greater than or equal to (>=), or less than or equal to

(<=).

Operator	Use	Description
>	op1 > op2	op1 is greater than op2
>=	op1 >= op2	op1 is greater than or equal to op2
<	op1 < op2	op1 is less than op2
<=	op1 <= op2	op1 is less than or equal to op2
==	op1 == op2	op1 and op2 are equal
!=	op1 != op2	op1 and op2 are not equal

Assignment operators

The assignment operator is expressed with an equal sign (=) and is used to assign a value to a variable. It can be used in conjunction with the array operator. The main limit on the assignment operator is that the expression and the variable must agree in type. In other words, you can't assign a string literal to an integer variable. Beyond this, you can use the assignment operator to link any expression that you'd like to a variable, giving you a lot of flexibility.

There are also a selection of compound assignment operators that will evaluate an expression and then assign the results to a variable in a single step. As with standard assignment operators, the variable and expression have to match types. Using compound assignment operators will save you both space in your code and time spent writing it, and can simplify the writing of your programs.

When you create a compound assignment operator, the equals sign always comes second, with the additional operator directly in front of it. The operations you can use for this include addition (+=), subtraction (-=), division (/=), remainder (%=), exclusive OR (^=), inclusive OR (|=), left shift (<<=), right shift (>>=), and unsigned right shift (>>>=).

Operator	Purpose	Example	Equivalent
+=	Addition	x += 2	x = x + 2
-=	Subtraction	x -= 2	x = x - 2
/=	Division	x /= 2	x = x / 2
*=	Multiplication	x *= 2	x = x * 2
%=	Modulus	x %= 2	x = x % 2

Equality operators

Looking at the equality operators, you could easily mistake them for compound assignment operators. Like the above, they involve a combination of symbols where the second is an equal sign, but despite this similarity in appearances, these operators have a much different function.

The equality operators are comparative in nature. They can determine whether two statements are equal (==) or determine whether they're inequal (!=). The result that's returned isn't a numerical value but rather a true or false, similar to conditional operators (more on those below).

Cast operators

When you want to work with assignment operators, you need the two values to be of the same type. If they're not, you can change the type of an object using a cast operator, denoted with the word "type" in the code and it'll convert the object in question from one type to another. It does have its limitations; you can convert from one primitive type to another or from one reference type to another, but you can't convert from a primitive to a reference type or vice versa.

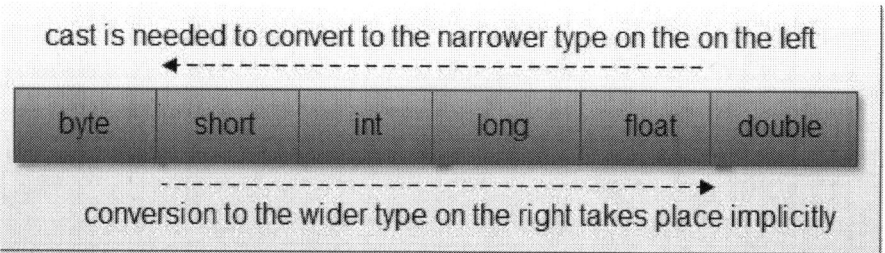

cast is needed to convert to the narrower type on the on the left

| byte | short | int | long | float | double |

conversion to the wider type on the right takes place implicitly

Array index operators

This has come up multiple times before in this book, though it wasn't referred to as an operator in those contexts. The array index operator is the set of brackets ([]) used to access an array element by providing its position in the array. The brackets are placed after the variable's name; since it only affects a single variable at a time, the array index is a unary operator.

You can also add the suffix .length to the end of an array. To reference the temps example from earlier in the book, you would access the length of the array by typing "temps.length" and would get a value back that indicates how many elements are assigned to the array.

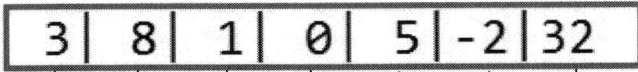

Object creation operators

Typing the word "new" at the start of a statement creates a new object from a class. You can follow this operator with the type of object, then put the object within parentheses following it. For example, the line "new String("XYZ")" will create that string.

You can also use this to create an array. You'll follow the same basic format as with the object creation above, except using brackets instead of parentheses after the identification of the type. The statement "new int[5]" creates a vector of five integers, while "new double[5][5]" would create a 5X5 table of double precision floating point values.

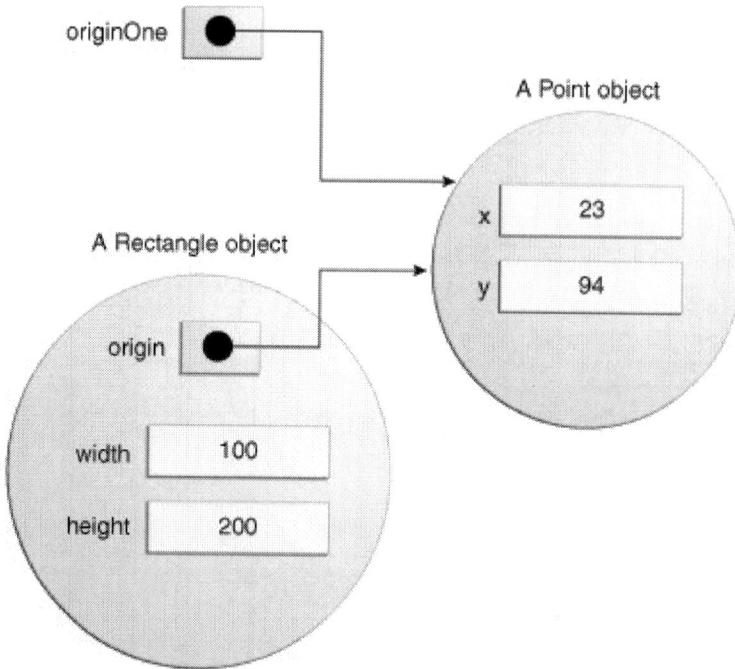

originOne

A Point object

x 23

y 94

A Rectangle object

origin

width 100

height 200

Bitwise operators

There are a number of commands in Java that allow you to alter the binary values of your objects in various ways. Included among these are three different shift options, complement, and three conditional operators.

The signed left shift operator (<<) is a unary operation that shifts a bit pattern to the left, while the signed right shift operator (>>) shifts it to the right. The unsigned right shift operator (>>>) always puts a zero in the left-most position, which is different than the signed right shift, where it depends on what was there before.

The bitwise complement operator (~) is a unary operation that flips the binary of the value in question, turning ones into zeros and vice versa. You can also perform a bitwise AND (bitwise &), a bitwise exclusive OR (bitwise ^), or a bitwise inclusive OR (bitwise |).

Conditional operators

These last two groups of operators are for evaluating what are called Boolean expressions, which are statements that, when evaluated, return a true or false result. Like comparative operators, these are very helpful in sending your program down different potential paths, especially when paired with if/then statements (covered in the next chapter)

	Operator	Type
unary operator →	++, --	Unary operator
	+, -, *, /, %	Arithmetic perator
	<, <=, >, >=, ==, !=	Relational operator
Binary operator	&&, ||, !	Logical operator
	&, |, <<, >>, ~, ^	Bitwise operator
	=, +=, -=, *=, /=, %=	Assignment operator
Ternary operator →	?: Tutorial4us.com	Ternary or conditional operator

There are three different conditional operations. The first, conditional, is a question mark and colon around two variables, as in ?A:B. This will return either A or B, depending on the outcome of a Boolean expression, often an expression in parentheses before the question mark. For example, if the whole statement read "(A>B)?A:B;" then it would return A if the statement is true and B if it's false.

Conditional AND (&&) also evaluates two values, but gives you a response of true only if both of them are true; if even one is false, the return is false. Conditional OR (||) evaluates two values and gives you a response of true if either response is true, only giving you a false response if both are false.

Logical operators

You could think of logical operators as the Boolean equivalent of the bitwise operators described above. The logical complement operator (!) flips the value of the operand (true to false and vice versa) in a similar way to how the bitwise complement flips the binary code.

There are also three conditional logical operators. The logical AND (&) and logical inclusive OR (|) have the same parameters as the AND and OR operators described above. The logical exclusive OR (^) returns a result of true only if one of the operands is true and the other is false.

Chapter 7 – Statements

The final building block of your Java code is the part that brings it all together and makes it work the way you want it to. When an application reads your code it evaluates it in discrete segments, each of which accomplishes a single discernible task. These segments are known as statements.

A simple statement takes up one line of code, gives stand-alone instructions for completing a single task, and is always terminated by a semi-colon. You can also combine simple statements together into compound statements. Denoted by braces, compound statements are also known as code blocks.

The variable statements described previously are essentially a form of simple statement, as are the assignment operators and object creation operators from the last chapter. You can also write decision statements, which order the program to choose between two or more paths of execution.

Some of the most common and useful statements are explored in the chapter below. By combining these together you can link all your values, expressions, and other objects into a truly valuable program.

If statements

The simplest decision statement is the "if" statement. You can use it to evaluate any Boolean expression, then execute another statement when the value is returned as true. Let's say you wanted to create a program that tells you whether someone is of legal drinking age. You could write a block of code like this:

```
{

if (age >=21)

    System.out.print1N("This individual is of legal drinking age")

}
```

The value of age could be set by typing it in, or extrapolated based on the current date and the individual's birthdate. When the age is over 21, the screen will display that the individual can legally drink.

```
public static void main(String[] args) {

    int user = 21;

    if (user <= 18) {
        System.out.println("User is 18 or younger");
    }
    else if (user > 18 && user < 40) {
        System.out.println("User is between 19 and 39");
    }

    else {
        System.out.println("User is older than 40");
    }
}
```

For a more advanced version of this process, you can use an if-then statement. This is a two-part piece of code that also gives the system instructions for a return of false. To use the same example as the one above, the code may look like this:

{

if (age >=21)

System.out.printin("This individual is of legal drinking age")

else

System.out.printin("This individual is " + (21-age) + "years under-age")

}

You can chain multiple if-else statements together, as well, using them to lead the program through a series of decisions based on information inputted by the user or generated by another section of the code.

Switch statements

The switch Statement

- Switch with default case:

```
switch (option){
    case 'A':
        aCount++;
        break;
    case 'B':
        bCount++;
        break;
    case 'C':
        cCount++;
        break;
    default:
        otherCount++;
        break;
}
```

If you want to choose from among several different execution paths, a switch statement is a more efficient means of doing so than a string of if-else statements. You can activate a switch statement with the method "switch ()" and the code that followed would look something like this:

switch (expression)

{

    case value1: statement1 [break;]

    case value 2: statement2 [break;]

    case value 3: statement 3[break;]

    [default: statement]

)

You can add as many cases as you want, each of which will execute a specific statement. The final line tells the system what to return by default, in the case that no other options return true.

Iteration statements

Also known as loop statements, these are designed to repeatedly execute a given statement. You can either write it to run for a specific number of times or to continue running until a specific condition is met. There are three main sub-categories of iteration sequences: for, while, and do-while statements.

For statements execute another statement for a specified number of iterations. The basic code looks like this:

{

```
for ([initialize]; [test]; [update])

    statement

}
```

As far as what those three words mean in the brackets, those are ways that you can customize your "for" statement. Initialize gives you a chance to list variable declarations or assignments that will be affected by the statement. Test allows you to enter a Boolean expression which will determine how long the loop executes; the iterations will continue so long as this remains true. Update lets you add a list of expressions, separated by commas, which modify the loop. All three of these modifications are optional; if you choose not to use them, remove the bracketed portion, but leave the parentheses and semi-colons.

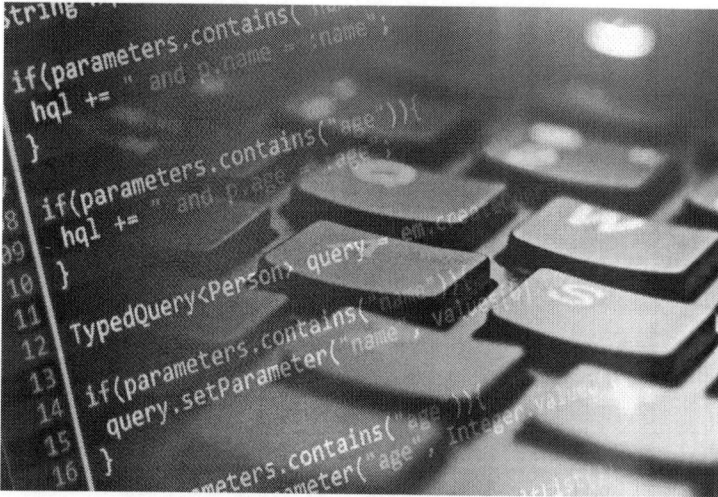

While statements repeatedly execute a different statement so long as the condition of a designated Boolean expression remains true. The basic code looks like this:

```
{
while (expression)
    statement
}
```

Do-while statements are a variation of the while statement, coded like this:

```
{
do
    statement
while (expression)
}
```

Break and continue

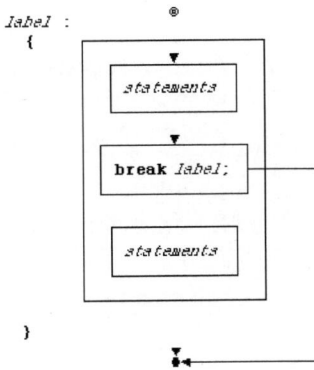

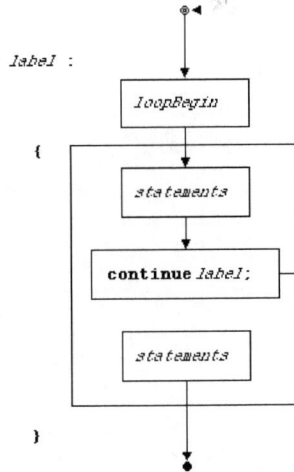

In the example code for the switch statement above there was the word "break" at the end of each case. This is a command that terminates any switch, for, or while statement, transferring the execution instead to the next statement in the code. It is essentially a way to move the progress of your program forward. If you want to break a designated statement, you can also use a labeled break by typing "break label;" and inserting the name of the statement for the word "label."

If you want to terminate an iteration sequence, you can also use the "continue" command. This tells Java to skip over the rest of the iterations in the sequence and move on to the next statement. Similar to breaks, you can label a continue command to end a specific iteration statement.

Chapter 8 – Using Comments

Not everything you write into your source code has to be something you intend the JVM to execute. Even for an experienced user, long sequences of code can be difficult to parse through. If you're just learning how to code, going back through to find errors can be especially frustrating.

The comments feature is not a required part of your coding, but using it can let you leave notes for yourself with information like what a specific line of code is meant to accomplish, or which update the code was introduced in. When you do want to fix or change something, having these notes in the code will make it a lot easier for you to figure out what you were doing the first time around.

The compiler ignores comments, so you don't have to worry about them messing with your program. They are simply there for the purpose of the user. Depending on how much information you need to impart, there are two different formats you can use for single line or multi-line comments.

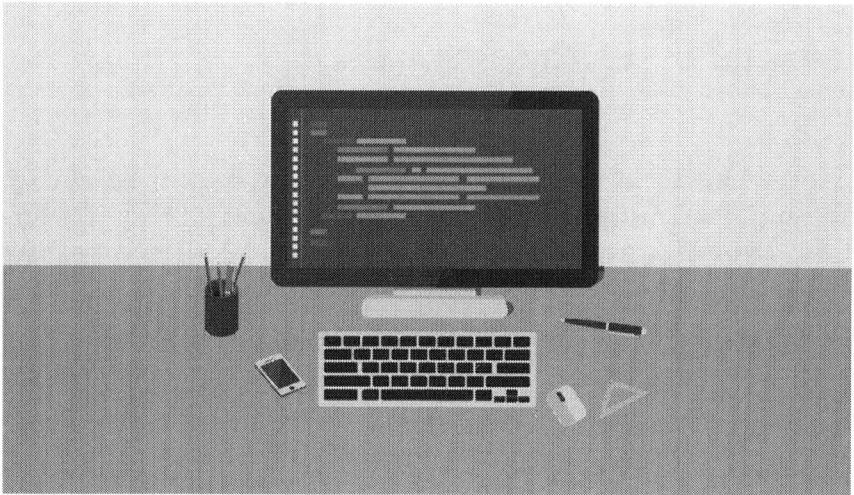

For a single line comment, type two slashes (//) followed by your comment. The compiler will ignore everything that follows on that line after the slashes. Make sure this kind of comment remains on one line; anything on the next will be read as if it's normal code. This kind of comment is the best for making simple statements and definitions, or short statements of intent behind specific lines of code.

If you need to go into more detail, use a multi-line comment. Start it with /* and end it with */. Everything that falls in between those symbols will be ignored by the compiler. This is most often used to describe entire sections of code, or to explain multiple statements at once.

The multi-line comment is also helpful in editing your code. Say you want to see how it would run without a specific section but you don't want to delete the code entirely. Since the compiler ignores whatever's between the comment symbols, you can simply place those around the code in question.

Javadoc comments

There is a special kind of comment that does affect what the end user sees, though it doesn't have any impact on the code. A Javadoc comment is still ignored by the compiler but is processed into an HTML-based documentation, index.html, which contains important information about the program.

A Javadoc comment ends the same way as a normal multi-line comment (*/) but it begins differently, with two asterisks after the slash (/**). Javadoc comments use tags that start with an at symbol (@) and clue the compiler in to specific information that's to follow.

```
/**
 * @author Sarah Smith
 *
 * The Inventory class contains the amounts of all the
 * inventory in the CoffeeMaker system.  The types of
 * inventory in the system include coffee, milk, sugar
 * and chocolate.
 */
public class Inventory {

    //Inventory code here

}
```

The author is identified with the tag @author. If you have a see also reference, you can type @see. Other information is about the program itself, like @param that describes the method parameters or @return that tells what kind of value the method returns. You can also use @throws to document exceptions. Finally, there are some tags you can use for specific segments of code. If you have source code that shouldn't be used anymore, you can tag it @deprecated. You can also document which software release the code originated in by tagging it @since, especially helpful with programs that undergo multiple upgrades.

Conclusion

The preceding chapters have introduced you to all the pieces you'll need in your Java programmer's toolbox. You know about expressions, variables, and objects, what a Boolean expression is, and how to generate an array. By forming operators and expressions into statements, you create the lines of code that make your program function.

This knowledge is what you'll need to write your first code—but that doesn't mean you're quite ready to pound out your first program just like that. The best next step to take is to find some source code attached to functioning programs that you can examine and fiddle with, identifying the key components of the code using the knowledge you gained in this book and watching how they interact.

An exception can be frustrating in the moment when you're writing code, but never underestimate the educational potential of these error messages. They point you to exactly where in the code the error occurred and help you figure out just what went wrong. Pay attention to these moments, and use them to continue to refine your understanding of the

Java language.

Start with a few small programs that do easily identifiable things. You could write a program that rolls a set of dice, for example, or one that tosses a coin. Writing programs such as these will help you get into the rhythm of writing code, paving the way to more complex code.

As with other languages—human or programming—practice is the key. Write a bit of code during your downtime to hone your skills and start internalizing the different actions and words associated with them. If you keep at it, you'll be writing your own applications in no time.

Thank you for reading. I hope you enjoy it. I ask you to leave your honest feedback.

53212568R00043